Ergonomics In Design

Engineering Anthropometry and Workstation Design

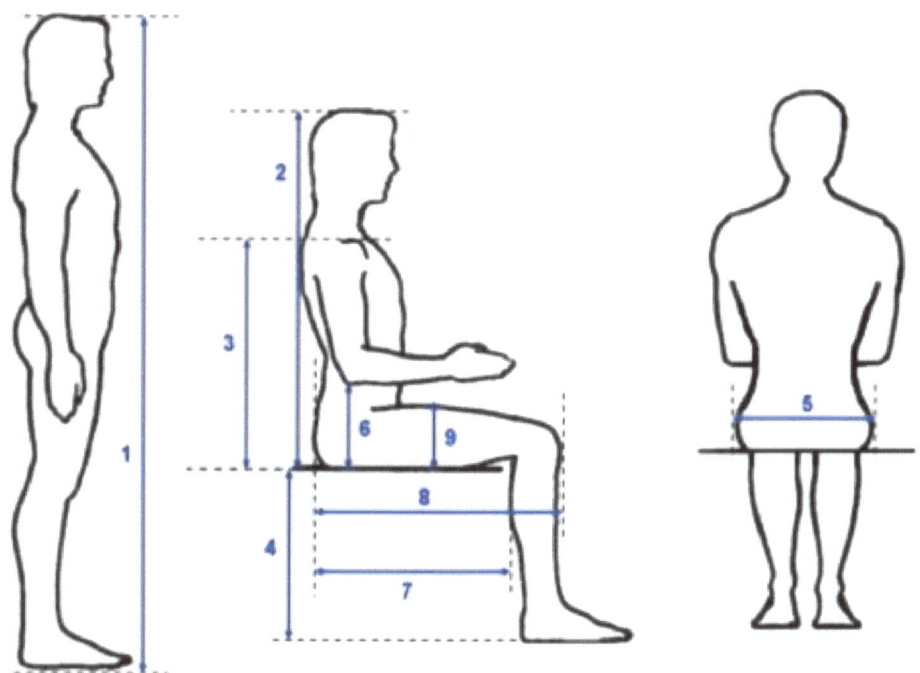

**Engr. MD Nursyazwi Mohammad
Greanna Friva Jainal**

This e-book was written by

Engr. MD Nursyazwi Bin Mohammad
BEng. (Hons.) Manufacturing Engineering (Design)

And edited by

Greanna Friva Binti Jainal
BA. (Hons.) Education and Science (Chemistry)

This work is licensed under a Creative Commons Attribution-NonCommercial-ShareAlike 3.0 Unported License.

Ergonomics In Design: Engineering Anthropometry and Workstation Design. Copyright © 2013 by Engr. MD Nursyazwi and Greanna Friva Jainal. All rights reserved worldwide. No part of this publication may be replicated, redistributed, or given away in any form without the prior written consent of the author/publisher or the terms relayed to you herein.

Engr. MD Nursyazwi Mohammad
Greanna Friva Jainal

Vi & Ci Associates Official Website

Table of Contents

Table of Contents 2
Meet the Author and the Editor 6
Foreword 7
Objective 8
Chapter 1: Introduction 9
    1.1: Age Variability 9
    1.2: Sex Variability 10
    1.3: Racial and Ethnic Group Variability 11
    1.4: Occupational Variability 11
    1.5: Generational or Secular Variability 11
    1.6: Transient Diurnal Variability 12
Chapter 2: Statistical Analysis 13
    2.1: Normal Distribution 13
    2.2: Mean 14
    2.3: Standard Deviation 14
    2.4: Percentiles 15
Chapter 3: Anthropometric Data: Measurement Devices and Methods 16
    3.1: Head Measurement 16
    3.2: Chest, buttock-popliteal length and acromial height (sitting) measurement by using Anthropometer 17
    3.3: Using measuring tape to measure stature height and waist 17
Chapter 4: Use of Anthropometric Data in Design 18
    4.1: Determine the user population 18
    4.2: Determine the relevant body dimensions 18
    4.3: Determine the percentage of the population to be accommodated 18
    4.4: Determine the percentile value of the selected anthropometric dimensions 18
    4.5: Make necessary design modification to the data from anthropometric tables 18
    4.6: Use mock-ups or simulators to test the design 18
Chapter 5: General Principles for Workspace Design 19
    5.1: Clearance requirement of the largest users 19
    5.2: Reach requirements of the smallest users 19
    5.3: Special requirements of maintenance people 20
    5.4: Adjustability requirements 21
    5.5: Visibility and normal line of sight 22
    5.6: Component arrangement 23
Chapter 6: Workstation Design for Standing and Seated 25
    6.1: Work Surface Height 25
    6.2: Standing position 26
    6.3: Sitting position 26
    6.4: Adjustable workstation design 27
    6.5: Working Envelope 28

Citation 29
Thank You 31

Meet the Author and the Editor

MEET THE AUTHOR AND EDITOR

The Author

E-book author named Engr. MD Nursyazwi Bin Mohammad, was an entrepreneur in the field of building construction, supplies and services, known generically as Wannah Enterprise. If you are interested, you can visit his website at http://wannah.net/.

Author comes from Sandakan, Sabah, Malaysia, and is one of the natives in Sabah, Momogun. The writer has a Diploma in Mechanical Engineering (Manufacturing Technology) and Bachelor of Manufacturing Engineering (Manufacturing Design).

Authors have extensive experience in the industry as a Machinist, welder, technician and engineer in some of the leading companies in Malaysia. However, now, active in the business arena.

The Editor

This e-book editor named Greanna Friva Binti Jainal. Editor is a teacher and now serves as an educator at SMK Muhibbah, Sandakan.

Editor is a woman who hails from Sandakan, Sabah, Malaysia. Editor's start higher education in Labuan Matriculation College and then, resume studies in Universiti Malaysia Sabah in the field of Bachelor of Education and Science (Chemistry).

Editor very interested in writing despite busy work in educating students. Editor is also very active in the uniformed units and an Assistant District Commissioner in Sandakan for Scout units.

Foreword

This eBook is about Engineering Anthropometry and Workstation Design in fact related to ergonomics.

With this eBook price which is much cheaper but, this eBook, filled with solid content that you can use for your daily life or in your career field.

Ergonomics is not a new knowledge, or new studies.

History has shown that, ergonomics has been practiced by the Greeks and the word ergonomics was also taken from the Greek word itself.

"Ergo" means work and "Nomos" means natural laws.

If combined, we can define it as:

> *Ergonomics is the science of designing the job, equipment, and workplace to fit the worker. Proper ergonomic design is necessary to prevent repetitive strain injuries, which can develop over time and can lead to long-term disability.*

I also hope that you can learn what is available in this eBook and use whatever you have to continue to learn and to not waste energy, time and money.

Md Nursyazwi

Engr. MD Nursyazwi Bin Mohammad
BEng. (Hons.) Manufacturing Engineering (Design)

Objective

At the end of this eBook, you should be able to:

- Define the term of anthropometry;
- Classify human variability;
- Determine the role of statistical method in engineering anthropometry;
- Obtain general principles for workstation design;
- Design workstations for standing and seated.

Chapter 1: Introduction

Anthropometry is the **study** and **measurement** of **human body dimensions**. It comes from Greek's word: anthropos (human), merikos (measurement).

It provides the fundamental basis and quantitative data for matching the workstations and products to the worker/user. Anthropometric data are used to develop design guidelines for heights, clearance, grips, and reaches of workstations and products.

Example: workstation (seating and standing); machinery; corridors; emergency exits; etc.

Anthropometry contains the physical geometry, mass properties, and strength capabilities. Anthropometry data are also applied in the design of consumers products such as clothes, automobiles, furniture, hand tools, etc.

Product design on the basis of male anthropometric data would not be appropriate for many female consumers. Applying the data collected from one country to other regions is not suggested (design failure).

Example: Apply anthropometric data of Europe populations to design clothes for Malaysians.

Human variability and statistics help designers analyze variability of population and how to use the data (using statistical information).

There are six (6) human variability:

- Age Variability,
- Sex Variability,
- Racial and Ethnic Group Variability,
- Occupational Variability,
- Generational or Secular Variability and
- Transient Diurnal Variability

1.1: Age Variability

- The stature (height) is increased until age about 20 to 25 (Roche & Davila, 1972; VanCott & Kinkade, 1972).

- The stature starts to decrease after about age 35 to 40, and women more shrinkage than men (Trotter & Gleser, 1951; VanCott & Kinkade, 1972).
- Unlike stature, weight and chest circumference increase through age 60.
- Example of age variability:

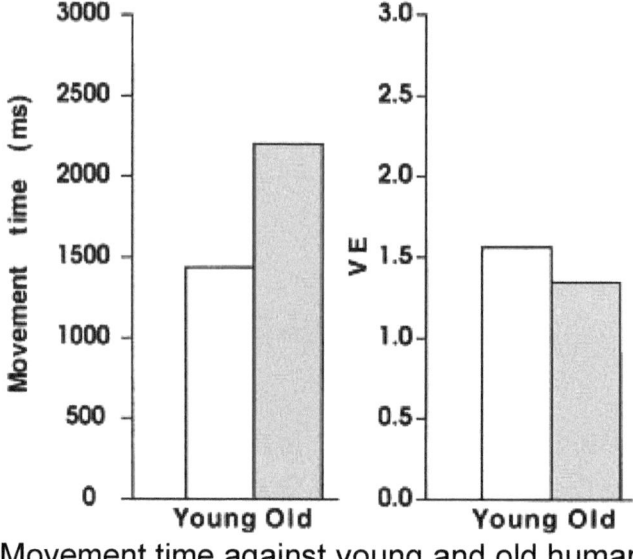

Movement time against young and old human

1.2: Sex Variability

- Adult men are, on average, taller and larger than adult women.
- However, at 12-year old, girls are taller and heavier than boys (on average).
- On average, adult female dimensions are about 92% of the adult male dimensions.
- Although adult men are larger than adult women, some dimensions are not! Eg: hip and thigh caused by skinfold thickness.

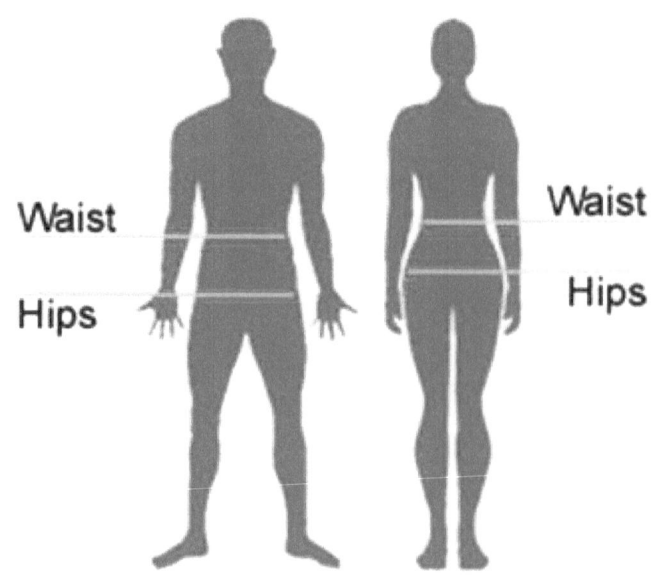

1.3: Racial and Ethnic Group Variability

- Black and white males in U.S: Blacks have longer arms and legs; shorter torsos than whites (Long & Churchill, 1965; Nasa, 1978).
- Japanese is shorter than American, but their average sitting height did not differ much.
- Ashby (1979) states that American = Europe > Frenchmen > Italians > Japanese > Thais > Vietnamese.

1.4: Occupational Variability

- Professional basketball players are much taller than most American males.
- Ballet dancers are thinner than rugby players.
- Truck drivers tend to be taller and heavier than average (Sanders, 1977).
- Coalminers have larger torso and arm circumferences (Ayoub et al., 1982).
- Factors determined the occupational variability are type and amount of physical activity, and special physical requirements.

1.5: Generational or Secular Variability

- Americans growth in stature about 1 cm per decade (Annis, 1978).
- Improved nutrition and living conditions are offered the growth.
- 22 body dimensions of male U.S army show a clear trend of growth (Griener and Gordon, 1990).

1.6: Transient Diurnal Variability
- Body weight varies by up to 1 kg per day because changes of water content in body (Kroemer, 1987).
- The stature may reduced up to 5 cm because of gravitational force on posture and the thickness of spinal disks.
- Chest circumference changes with the cycle of breathing.

Chapter 2: Statistical Analysis

In engineering design, the variabilities are analyzed using statistical distribution rather than a single value.

Normal distribution is commonly used because it approximates most anthropometric data quite accurately.

Normal distribution visualized as normal curve as "Symmetric, bell-shaped curve".

Mean is calculated as sum. of individual measurements divided by sample size.

2.1: Normal Distribution

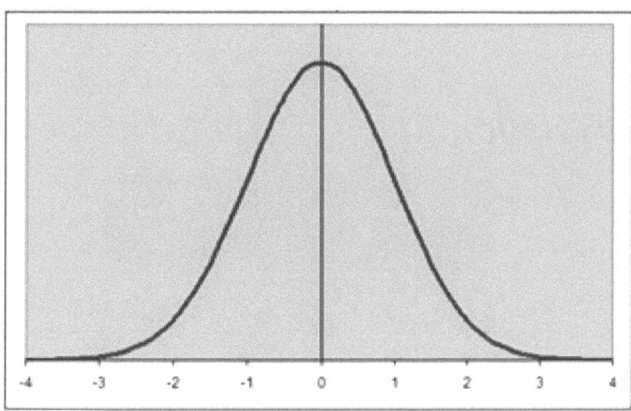

Figure 1 -- Standard Normal Distribution

1. Mean = sum. of measurement / sample size.
2. Symmetric, bell-shaped curve.

2.2: Mean

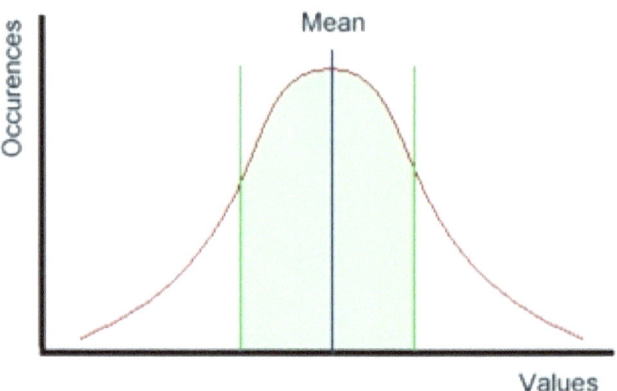

Mean is the sum of all the individual measurements divided by the number of measurements.

- It is a measure of central tendency.
- Thus, the value of the mean determines the position of the normal distribution along the x-axis.

2.3: Standard Deviation

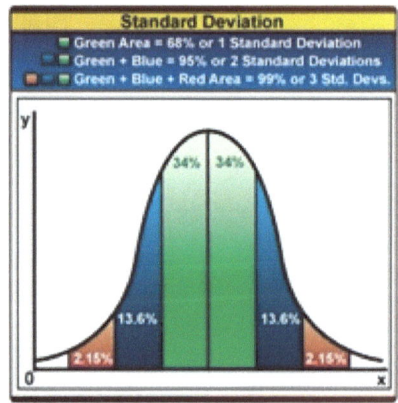

Standard Deviation is calculated using the difference between each individual measurement and the mean.

- Measure of the degree of dispersion in the normal distribution.
- The value of the standard deviation determines the shape of the normal distribution.
- Small value> close to the mean value.
- Large value> scattered more distantly from the mean.

2.4: Percentiles

- A percentile value represents the percentage of the population with a body dimension of a certain size or smaller.
- Percentile helps designers to estimate the percentage of a user population that will accommodated by a specific design.
- **Example**: if the width of a seat surface is designed using 50th percentile of hip breadth of U. S males, it can be estimated that about 50% of U. S males (those with narrower hips) can expect to have fully supported; whereas the other 50% (those with wider hips) cannot.
- For normal distribution, 50th percentile = mean of the distribution.

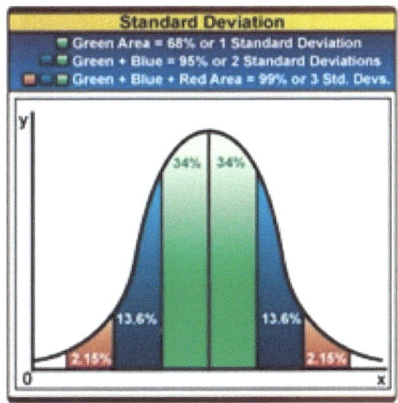

Percentile is calculated by:
$$X = M + F \times s$$

where:
X = percentile
M = mean (50th percentile value)
F = multiplication factor (refer Table 10.1)
s = standard deviation

Chapter 3: Anthropometric Data: Measurement Devices and Methods

- Measuring tape can be used to measure circumferences, contours, curvature as well as straight body parts.
- An anthropometer set has straight, graduated rod with one sliding and one fixed blade, caliper, hole board, and compass.
- Straight and graduated rods can be used to measure the distance of body segments.
- Caliper is used to measure distance between the tips of the two branches.
- A compass is used to measure short distance of body segments.
- Hole board is used to measure the diameter of round body parts; e.g. fingers.

3.1: Head Measurement

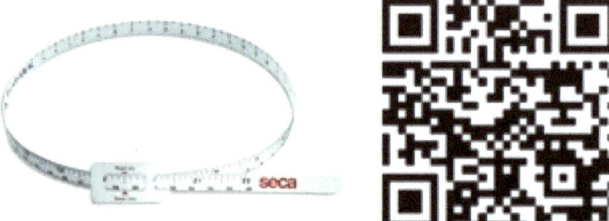

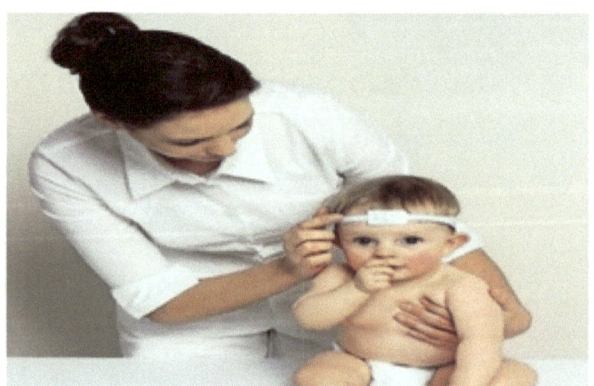

Click here for details

3.2: Chest, buttock-popliteal length and acromial height (sitting) measurement by using Anthropometer

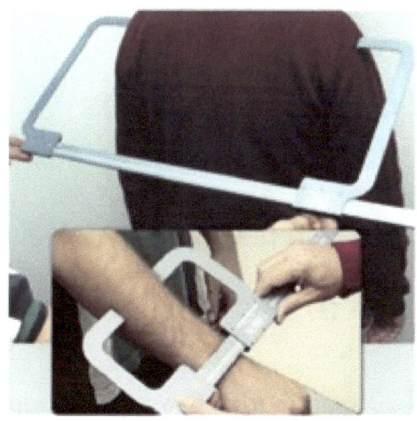

[Click here for details](#)

3.3: Using measuring tape to measure stature height and waist

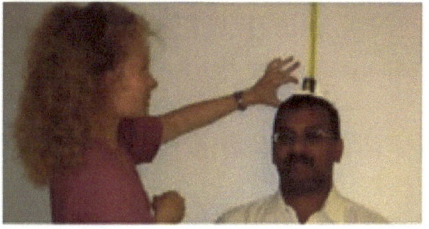

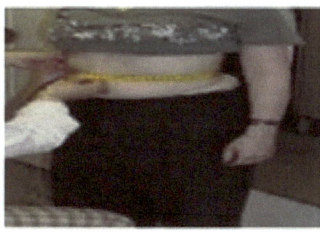

Click Here For More Details

Chapter 4: Use of Anthropometric Data in Design

4.1: Determine the user population
- Who will use the product or workstation?
- Considering age, gender, race, ethnic, military or civilian?, etc.

4.2: Determine the relevant body dimensions
- Which body parts are most important to design the products or workstations?
- Head size, knee height, thigh clearance, etc.
- For example: To design a helmet, which body dimension is most important?

4.3: Determine the percentage of the population to be accommodated
- Can the design accommodate 100 percent of the population?
- **For example**: design of a car seat may not accommodate to all drivers in the world.
- **Solution**: provide adjustable seat.

4.4: Determine the percentile value of the selected anthropometric dimensions
- Which percentile value of the relevant dimension to be used (5th, 50th, 95th)?
- Should the percentile value be selected from the male or female population?

4.5: Make necessary design modification to the data from anthropometric tables
- Adjustments are needed to accommodate the changes of variables such as clothing can change body size considerably, light shirt for the hot work environment is different from heavy jacket for cold work environment.
- **Working posture**: standing, seated, and standing/seated position.

4.6: Use mock-ups or simulators to test the design
- To obtain the interactions between users and products/ workstation.
- It can helps designers to make necessary correction at preliminary design.

Chapter 5: General Principles for Workspace Design

- Clearance requirement of the largest users
- Reach requirements of the smallest users
- Special requirements of maintenance people
- Adjustability requirements
- Visibility and normal line of sight
- Component arrangement

5.1: Clearance requirement of the largest users

Clearance is the space between the human and the equipments and workplace.

- **Example**: the height and width: if the passageways are designed too low and narrow, high and large worker may difficult to access.
- **Example**: If the machines are arranged too close, large worker cannot go through.

Good design of passageways provide sufficient clearance and clearly marked

5.2: Reach requirements of the smallest users

- A short worker unable to reach the feed lever

Solution: Provide a platform

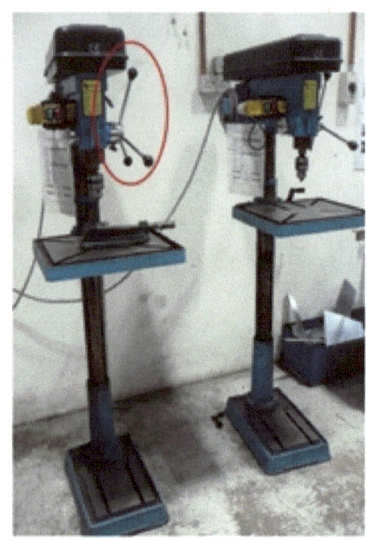

5.3: Special requirements of maintenance people
- A maintenance difficult to access the lower area of machine.

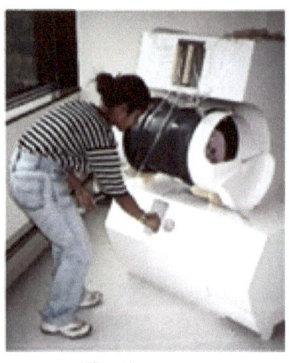 

Before   After

5.4: Adjustability requirements

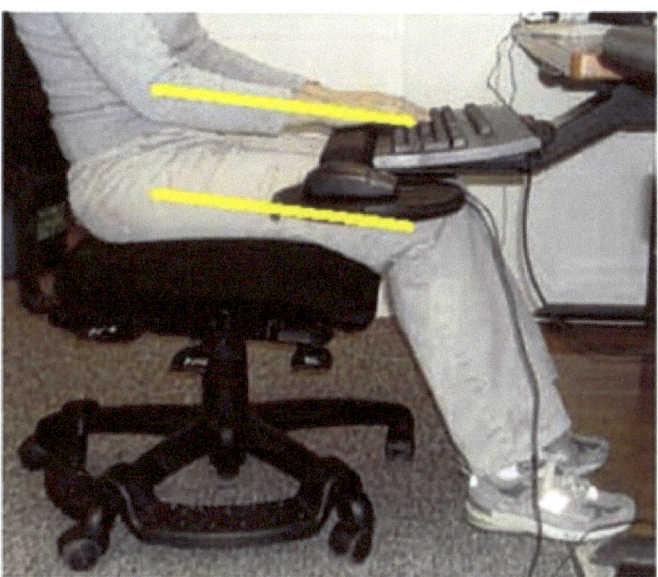

Tip 1: Position the keyboard so that your forearms are parallel to your thighs when your feet are flat on the floor.

Why?
- Helps maintain blood flow in the hands and arms.
- Decreases muscle strain and tension.

Laptop Computers

LAPTOP COMPUTERS ARE NOT RECOMMENDED AS PRIMARY COMPUTERS. IN THE OFFICE OR WHILE AT HOME, A DOCKING STATION IS RECOMMENDED TO PROVIDE ADJUSTABILITY WHICH WILL ENHANCE NEUTRAL POSTURES.

Are you using a laptop like this? – Don't!

Maintaining neutral postures will reduce stress and strain to the musculoskeletal system.

5.5: Visibility and normal line of sight

Viewing Angles and Distance

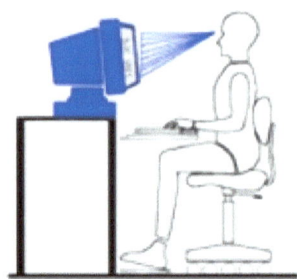

- Position monitor, keyboard, and chair in straight line
- Top of monitor screen at or slightly below the eye horizon
- Comfortable, non-straining distance to read monitor (typically 15-30 inches)
- Adjust brightness, contrast, and color to comfortable levels
- Avoid glare by positioning away from windows and direct lights
- Keep written work materials propped up near monitor screen

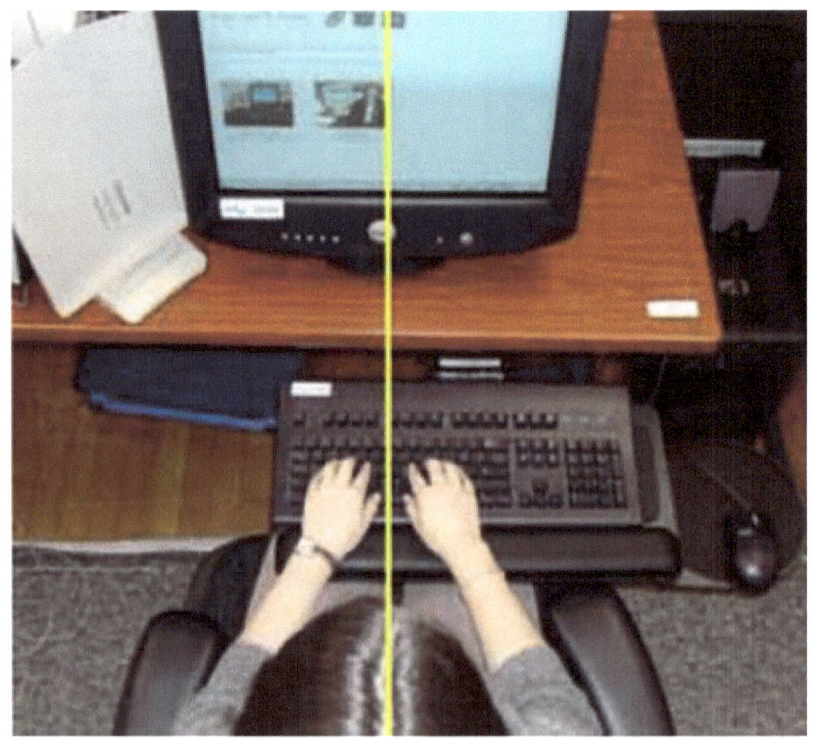

5.6: Component arrangement

Comfort arrangement   Discomfort arrangement

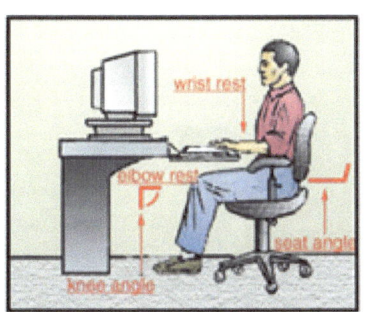

Chapter 6: Workstation Design for Standing and Seated

- **Standing workstation** = workers need more frequent movement, large work area, handle heavy objects, exert large forces with their hands.
- **Seated workstation** = long term duration jobs, allows better control of arm movement, provides stronger sense of balance and safety, improves blood circulation.
- **Sit/stand workstation** = workers can choose whether perform the job in standing or sitting working position.

6.1: Work Surface Height

- **Standing position** = the table height is 5 to 10 cm below elbow level.
- **Sitting position** = table height is at elbow level; except jobs require precise manipulation and great force application (Ayoub, 1973; Gradjean, 1988; Eastman Kodak Company, 1986).
- Adjustable work surface height is most prefer to accommodate short/tall workers.

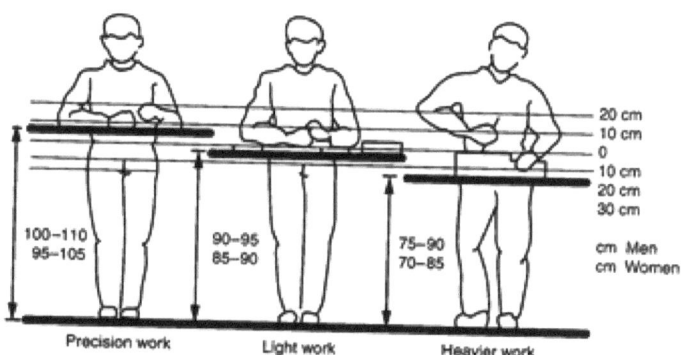

FIGURE 10.9
Recommended work surface height for standing work. The reference line (0 cm) is the height of the elbows above the floor. (Source: Grandjean, 1988, *Fitting the Task to the Man* [4th ed.]. London: Taylor and Francis.)

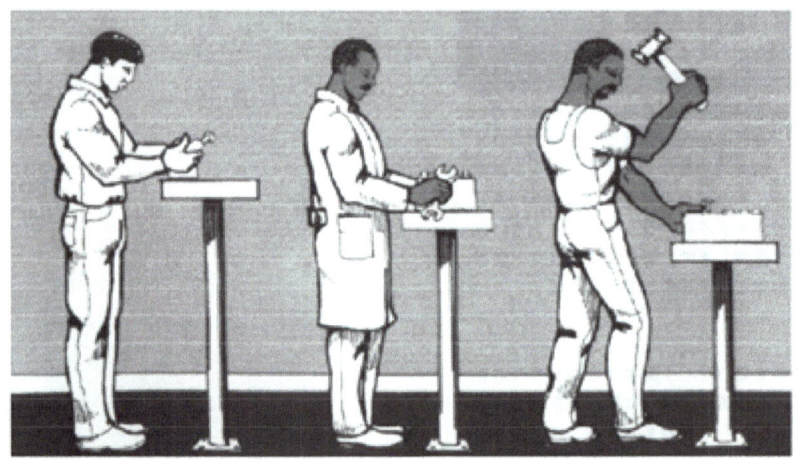

6.2: Standing position
The table height (conveyor) is 5 to 10 cm below elbow level.

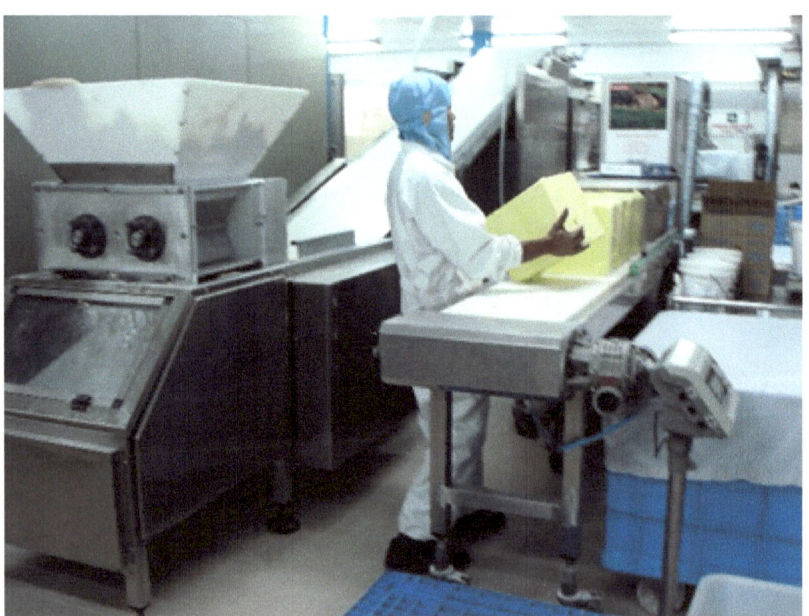

6.3: Sitting position
Table height is at elbow level

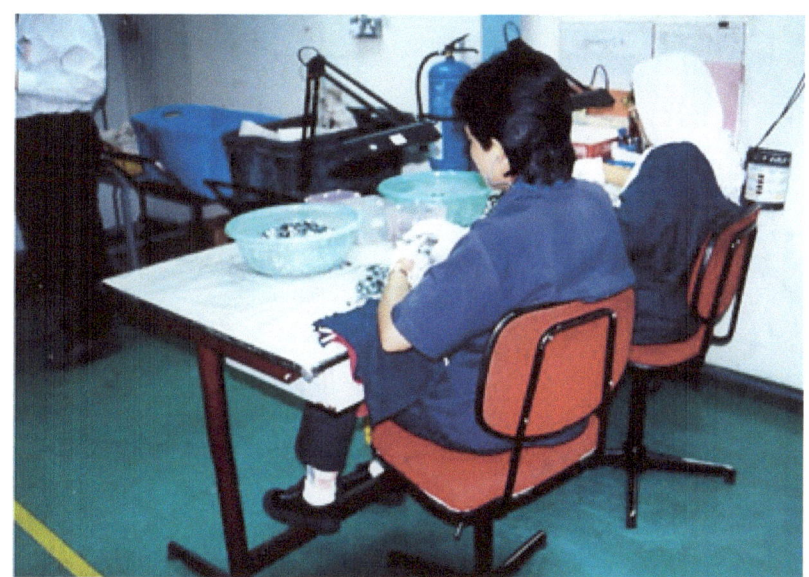

6.4: Adjustable workstation design

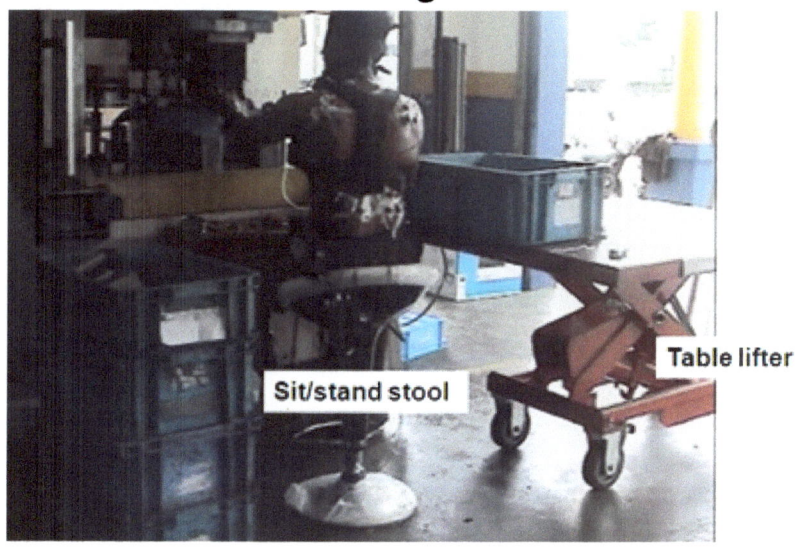

6.5: Working Envelope

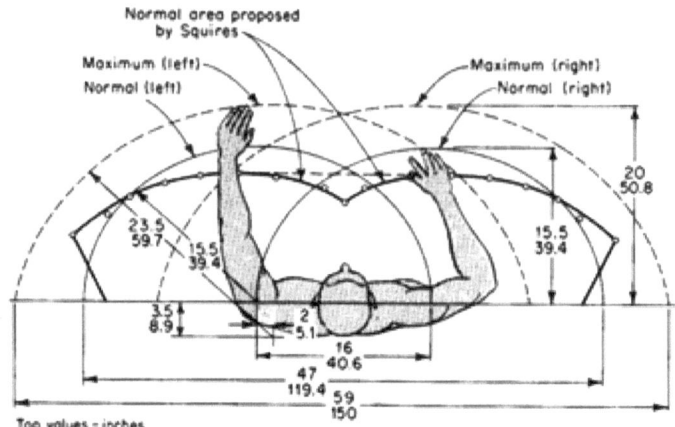

FIGURE 13-11
Dimensions (in inches and centimeters) of normal and maximum working areas in horizontal plane proposed by Barnes, with normal work area proposed by Squires superimposed to show differences. (*Source: Barnes, 1963; Squires, 1956.*)

Citation

Book

Jan Dul , (2008). Ergonomics for Beginners: A Quick Reference Guide, Third Edition. 3rd ed. America: CRC Press.

R.S. Bridger, (2008). Introduction to Ergonomics. 3rd ed. America: CRC Press.

Gordon Inkeles, (1994). Ergonomic Living : How to Create a User-Friendly Home & Office. 1st ed. America: Touchstone.

Online

Oklahoma State University (2011). What is Ergonomics?. [ONLINE] Available at: http://www.ehs.okstate.edu/modules/ergo/What.htm. [Last Accessed 2010].

Jan Dul (2009). Ergonomics for Beginners: A Quick Reference Guide, Third Edition . [ONLINE] Available at: http://amzn.to/1aX0cYQ. [Last Accessed 30 August, 2011].

R.S. Bridger (2009). Introduction to Ergonomics. [ONLINE] Available at: http://amzn.to/196ZOFD. [Last Accessed e.g. 31 August 11].

Thank You

This e-book was written by

Engr. MD Nursyazwi Bin Mohammad
BEng. (Hons.) Manufacturing Engineering (Design)

And edited by

Greanna Friva Binti Jainal
BA. (Hons.) Education and Science (Chemistry)

This work is licensed under a Creative Commons Attribution-NonCommercial-ShareAlike 3.0 Unported License.

www.ingramcontent.com/pod-product-compliance
Lightning Source LLC
Chambersburg PA
CBHW050418180526
45159CB00005B/2326